Y0-DFA-202

DISCARDED

REACHING FOR THE STARS

M.C. HAMMER
Top Rapper

by Rosemary Wallner

Published by Abdo & Daughters, 6535 Cecilia Circle, Edina, Minnesota 55439.

Library bound edition distributed by Rockbottom Books, Pentagon Tower, P.O. Box 36036, Minneapolis, MN 55435.

Copyright © 1991 by Abdo Consulting Group, Inc. Pentagon Tower, P.O. Box 36036, Minneapolis, MN 55435. International copyrights reserved in all countries. No part of this book may be reproduced in any form without written permission from the publishers. Printed in the United States.

Library of Congress Number: 91-073039   ISBN: 1-56239-054-6

Cover photo: Retna Photos
Inside photos: Ron Riesterer Photo: 8; Retna Photos: 4, 19, 21, 25, 28, 30; UPI/Bettmann: 13

Edited by Bob Italia

TABLE OF CONTENTS

An Awesome Entertainer..................................5

Growing Up in Oakland..................................6

The Struggle for Success..............................10

A Deal with Capitol Records...........................12

Please Hammer Don't Hurt 'Em..........................15

On Tour...18

Criticism and Awards..................................23

A Tireless Businessman................................26

M.C.'s Future Plans...................................29

M.C. Hammer's Address.................................32

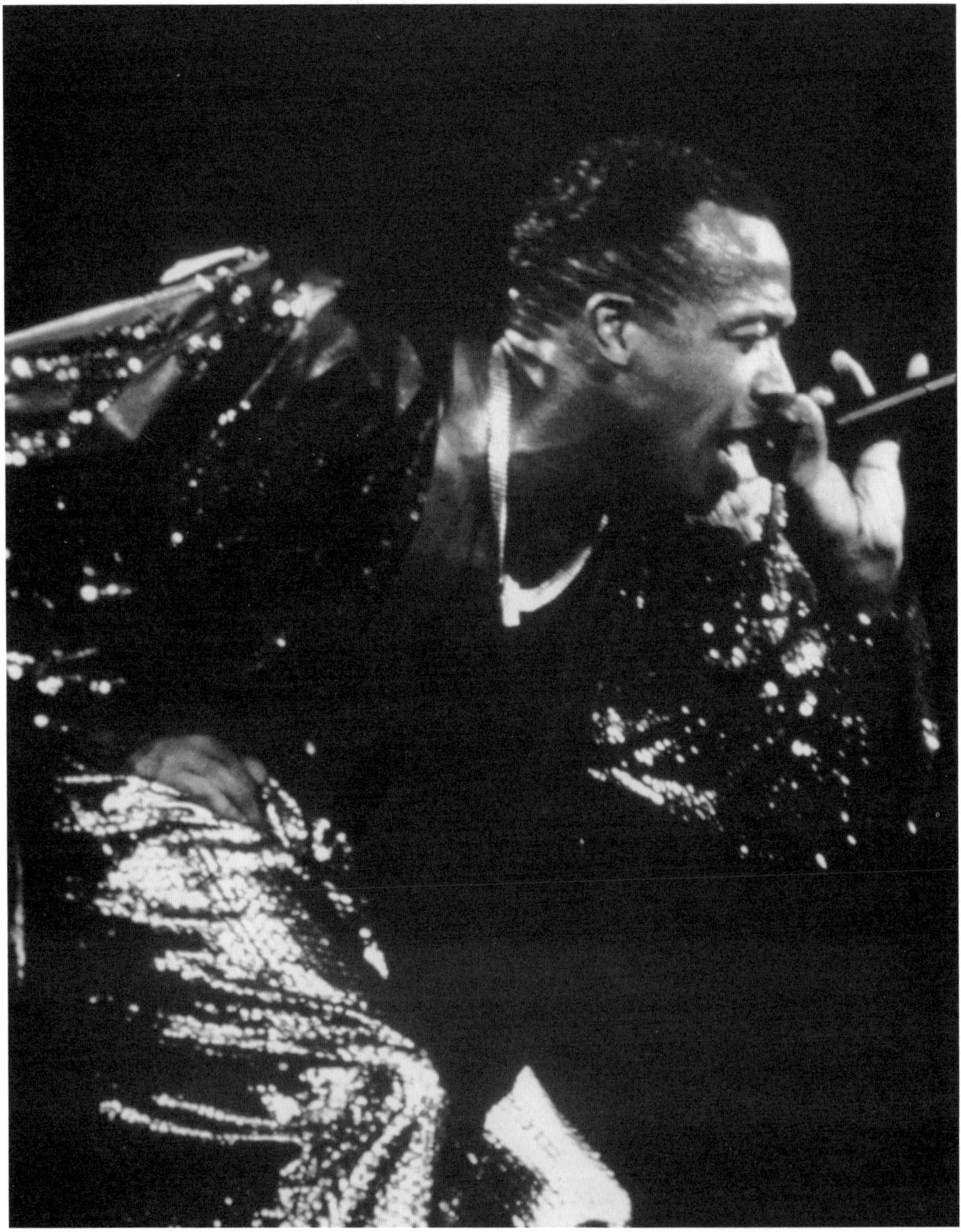

AN AWESOME ENTERTAINER

M.C. Hammer is no ordinary rap singer and no ordinary performer. "I'm totally different, a whole different ball game," he said when describing himself. With his harem pants, up-to-date glasses, and show-stopping dance moves, it's easy to see why Hammer is so popular. His second album *Please Hammer Don't Hurt 'Em* is only the third rap album in history to land at the top of the pop charts. Since its debut in 1990, it has sold nearly eight million copies.

His warm, slightly hoarse voice combined with his liquid-legged dancing sell out his concerts wherever he performs. After one concert in Maryland, a magazine reviewer wrote, "Hammer's dancing acrobatics were astounding as he slid, leaped, crawled, twisted, and shook through roof-raising numbers." His steps have inspired new dance crazes and led Los Angeles to declare December 7, 1990, as M.C. Hammer Day.

Hammer watched the success of other performers such as James Brown, the Jacksons, and Prince.

Whether he's dancing or singing, M.C. Hammer works tirelessly to perfect each performance.

He's taken their styles and rolled them all into a unique, rapping, pulsating sound.

"I present to America a very exciting package as an entertainer," explained Hammer. "My show crosses all barriers and boundaries."

As one of the most popular performers of the early 1990s, M.C. Hammer can hardly wait for what's ahead.

GROWING UP IN OAKLAND

M.C. Hammer was born Stanley Kirk Burrell in 1963 in Oakland, California. His father worked as the manager of a legalized gambling club and his mother was a secretary. His parents divorced when Burrell was five years old. "We were definitely poor," he remembered. "We had three bedrooms and six children living together at one time."

As a boy, Burrell's hobbies were writing poetry, dancing, and playing baseball. To make some money, he and a friend collected stray baseballs at the nearby Oakland Coliseum and sold them for $2 each.

Other times, the boys danced in the parking lot. It was here that Charles O. Finley, the owner of the Oakland Athletics baseball club, discovered Burrell.

"Charlie saw me doing James Brown splits in the parking lot when I was eleven years old and he thought it was funny," said Hammer. "It impressed him, and he put me to work in the office running errands." Young Burrell spent seven years with the team. He traveled around the country with the ball players during spring breaks. For his duties as batboy and errand runner, Burrell earned $7.50 per game.

During one of the out-of-town trips in the early 1970s, a visiting baseball player noticed how much Burrell looked like baseball great "Hammerin' " Hank Aaron. The players began to call Burrell "Little Hammer." (Later, when Hammer began to dance at various rap clubs, he added the "M.C." which stands for "master of ceremonies.")

Hammer enjoyed his time with the Oakland Athletics. Whenever Finley went out of town, Hammer called him from the owner's box during a game and described the game play-by-play. He even tried his hand at broadcasting and created his own radio show called "HammerTalk." The show lasted until the station's general manager returned from his vacation and ordered Hammer off the air.

Twelve-year-old M. C. Hammer poses with his look-alike "Hammerin' "Hank Aaron.

Finley encouraged young Hammer to be whatever he wanted to be. "Charlie always said to me, 'Hammer, I can't promise that you're going to be a professional ball player, but I can say you'll make a positive contribution to your race.' And he said, 'You're going to be somebody of importance.'" Hammer thought he could become important by playing professional baseball. He had played in high school and was ready for a shot at the pros, but his tryout with the San Francisco Giants didn't work out.

Disappointed, Hammer enrolled at East Los Angeles College. After a year, he dropped out of college because he had trouble paying his bills. Back home in Oakland, he found that many of his friends had turned to selling drugs in order to make money.

"Everyone who had any pocket change had it because they were dealing a little drugs," said Hammer. He considered joining his friends but then realized he liked his clean, positive life. Instead of selling drugs, Hammer took a different route. "I went to the nearest recruiting office," he explained, "got all fifty questions on the test right and joined the Navy." For the next three years Hammer was stationed at a base in California and even spent some time serving in Japan. When his enlistment was over, Hammer was ready to follow his other dreams: dancing, singing, and performing.

THE STRUGGLE FOR SUCCESS

In 1982, Hammer returned to Oakland from the Navy and became a born-again Christian. He studied the Bible six days a week. He even formed a religious rap duo called the Holy Ghost Boys. Hammer persuaded two record companies to release a Holy Ghost Boys album, but eventually abandoned the project.

Hammer decided that the best way to make records was to start his own record company. In 1987, he borrowed money for this new venture from Dwayne Murphy and Mike Davis, two Oakland A's baseball players. Hammer was not a schooled singer and he couldn't play an instrument, but he was determined.

Davis remembered how excited Hammer was with his new project. "He'd play tapes, do some dancing," Davis recalled. "He said he needed someone to believe in him." Murphy and Davis each gave Hammer $20,000. In return, Hammer promised that the two men would receive ten percent of all his royalties.

With the money, Hammer founded Bust It Productions and recorded his first song "Ring 'Em."

Because he didn't have the money for expensive recording equipment, Hammer and his small crew recorded the song out of a friend's apartment closet. "The worst part," remembered the album's producer, "was that every night after ten o'clock, an old woman who lived in the apartment upstairs would bang on my ceiling with a broom while we were trying to record."

"Ring 'Em" was included in Bust It Production's first album titled *Feel My Power*. Hammer sold copies of the record out of the trunk of his car. Hammer's wife, Stephanie, sent promotional copies to DJs and record studios. The album sold about 60,000 copies.

At the same time, Hammer put together a stage show and asked his boyhood friends to act as dancers, singers, and DJs. The group worked day and night to make every dance step perfect. "We rehearsed thirteen to fourteen hours a day, seven days a week," said one dancer. "We even worked on Christmas Eve."

Although Hammer was becoming well-known, he knew he needed the backing of a major record company. And that's what he went after next.

A DEAL WITH CAPITOL RECORDS

In 1986, Run D.M.C. hit the charts with their rap single. It was obvious to many people in the music business that rap was swiftly becoming popular with many different types of people.

"It didn't take a rocket scientist to figure out rap was coming on strong," said Step Johnson, senior vice president of Capitol Record's black music division. Johnson knew that Capitol wanted to enter the rap market, but he was not sure how to do it.

In May 1988, Hammer floored Johnson and other Capitol executives with his dance steps and marketing ideas. "Hammer was an entertainer as well as an extremely astute businessman," said Johnson.

In his presentation to Capitol, Hammer told the executives he had two roles. "One [role] is artistic, one is business," he said. "They both make me happy in their proper perspective." Capitol liked Hammer's ideas and offered him a $750,000 advance.

Hammer and Capitol's first project was to rework *Feel My Power*. They added four new tunes and renamed the album *Let's Get It Started*.

The ever-popular M.C. Hammer waves to his fans.

The new album sold more than a million-and-a-half copies and had three Top Ten singles.

"Hammer is a pure entertainer," complimented Johnson. "He's done a lot to deliver his music to a segment of the population that otherwise never would have heard of rap." Many people say Hammer brought rap out of inner-city neighborhoods and introduced it to mainstream pop music.

Despite his success as a rap artist, Hammer remains sensitive about the label. "Some people have the wrong idea about rap," he explained. "I don't put on a riot-type of show. I just want to inspire people to have a good time and dance and feel free.

"I would rather be called an entertainer than a rap artist," he added. "I rap. I sing, dance. I entertain."

In late 1988 Hammer began work on his second album. Although *Let's Get It Started* had been a success, Hammer wanted different music on his next album. "I decided [the second album] would be more musical," he said. With his advance from Capitol, he equipped a tour bus with $50,000 worth of recording gear. As they toured the country promoting the first album, Hammer and his crew recorded new songs in the back of his tour bus. By the end of 1989, the second album was ready for the public.

PLEASE HAMMER DON'T HURT 'EM

Please Hammer Don't Hurt 'Em was released in February 1990. Within the first five months of its release, it sold five million copies. By mid-July 1990, an estimated half-million albums were being sold in a single week. The album first topped the charts in June. It stayed at the top of Billboard's pop chart for a record-breaking eighteen weeks. M.C. Hammer was suddenly a household word; his music was a hit with all ages.

Part of the success of the new album was due to a clever marketing idea cooked up by the sales department at Capitol Records.

During April 1990, mailings were sent to over 100,000 youths. Each mailing contained a cassette single of Hammer's rap version of the Jacksons's 1974 hit "Dancing Machine" and a letter signed by Hammer. The letter's messages were simple: Give his new record a chance and stay in school. "Nobody wins who quits on their future. Stay in school," Hammer wrote at the end of the letter.

"We were looking for a very specific direct-mail piece that would allow Hammer to give the message he wanted to," explained Lou Mann, Capitol's vice president of sales. "We would get the music out there in front of them and basically say to the kids, 'Hey listen, Hammer is here and he cares about you.'"

By far the most popular single on the *Please Hammer* album was "U Can't Touch This." Hammer based his lyrics off of a 1981 hit titled "Super Freak." The song's writer, Rick James, took legal action against Hammer when Hammer used his song without paying royalties. Hammer defended himself by saying he had every intention of paying James. Hammer explained that "right after I recorded the song I said, 'I have got to pay Rick for this.'"

The two performers managed to settle their differences out of court. Hammer told reporters that he paid "big bucks" to James. The two met for the first time in the summer of 1990 when Hammer performed in James' hometown of Buffalo, New York. Backstage, the two men promised that one day they would pool their talents and work on a project together.

Hammer summed up the incident by saying, "It's business. If you borrow someone's song you pay a certain royalty rate. You work out a business deal.

Rick James had a record. I paid him a certain percentage to do it."

Other songs on the *Please Hammer* album were also borrowed from other performers. The single "Have You Seen Her" was first recorded by the Chi-Lites in the 1970s. Hammer's "Help The Children" is based on Marvin Gaye's single "Mercy Mercy Me." Hammer credited these writers on his album and is also paying royalties.

Several songs on the album, including "Help The Children," make strong anti-drug statements. "Drugs is the problem I'm closest to," explained Hammer. "I grew up with it. I'm trying to create awareness among those who are *not* using drugs, and among those who *are* using drugs that their lives can be *shortened*. I'm not attacking drug dealers. I'm attacking the *problem* of drugs."

Along with his album, Hammer produced an anti-drug video titled *Please Hammer Don't Hurt 'Em: The Movie*. The video is about Hammer returning to his hometown and stopping the drug dealers.

ON TOUR

With the success of his second album, Hammer launched a 250-concert world tour that began in June 1990. He performed in the United States, Europe, Japan, Australia, and the Caribbean. For the tour, Hammer assembled fifteen dancers, twelve background singers, eight security men, seven musicians, three valets, and two DJs. The entire tour group flew from one destination to another in Hammer's private Boeing 727 jet.

As with everything Hammer does, his 1990-91 concerts were flashy and full of life, music, and fun. The stage set was designed to look like the rooftop of an inner-city building. Giant illuminated hammers swung down from the ceiling during certain songs. Over thirty dancers and singers appeared on stage in all kinds of extravagant costumes. And of course there was dancing. Hammer and his group broke new ground with their elaborate, high-voltage, fast-moving dance numbers.

"I like his moves," declared one thirteen-year-old Chicago fan. Another fan was impressed with Hammer's "clean-cut" image and "happy music."

For his 1990 tour, Hammer and his crew members put together an exciting show filled with rapping and plenty of dancing.

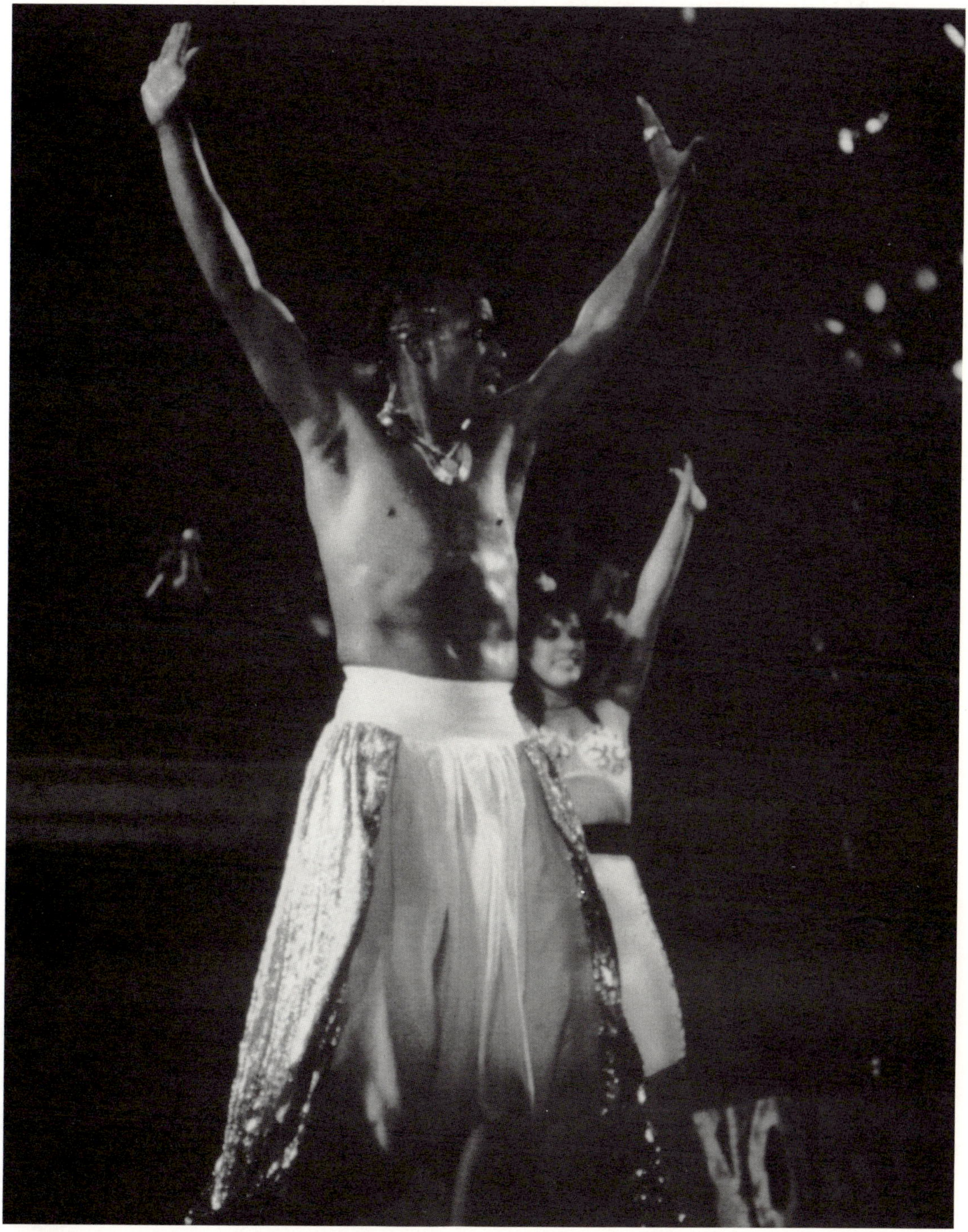

"My live shows are an event," boasted Hammer. "Anybody can come to see them. Even if they don't like rap music, they can come and say, 'Wow, I've just been to a show' . . . Everybody in my show dances great. We don't do that usual rap stuff of just walking back and forth."

During breaks between shows, Hammer dressed and acted a little more casually. He kept his glasses and baggy genie pants in the costume trunk and preferred to wear a one-piece bicycle outfit and cap with the flap turned up.

In his free time he attended baseball, football, and boxing events. (He still says the Oakland A's are his favorite baseball team.) Going to these events helped him relax, he said. He also loved to talk about the expensive classic cars he hopes to own one day. Throughout the tour he exercised faithfully and continued to create new dance steps for himself and his crew. Hammer insisted that all of his dancers be in top physical condition. Everyday, he and his tour group jogged four miles, lifted weights, and danced at least six hours.

Hammer was also strict when it came to the behavior of his tour group. Almost all members were told to return directly to their hotel rooms after each show and remain there for the rest of the evening.

During breaks between concerts, Hammer packs away his famous genie pants and wears a biker's cap, shirt, and pants.

One hundred dollar fines were given out to anyone who disobeyed this rule. Fines were also given out to tour members who missed dance steps on stage or who did not have luggage ready on travel days.

Some of the members were not so happy with these arrangements. "He (Hammer) wanted to have total control over everybody at all times," said Dontay Newman, a bodyguard who quit the tour.

But Hammer had a reason for his strictness. "We don't put curfews on you to control your life–just curfews that kind of help save your life," he explained. "Everybody is not twenty-five or thirty years old here. We've got eighteen-year-olds and nineteen-year-olds who we feel very responsible for."

Besides, added Hammer, "We have goals and to achieve those goals we must be a disciplined organization."

CRITICISM AND AWARDS

"Whenever you're on top, you're going to have controversy," stated Hammer in a December 1990 interview. And he's had plenty of controversy–and criticism.

Many critics say that Hammer is not a rapper, but only an exciting, tireless entertainer using rap as his medium. Others say that he is more of a dancer than a rapper.

Hammer's response? "People were ready for something different from the traditional rap style," he said. "The fact that ["U Can't Touch This"] has reached [number one] indicates the genre is growing."

Other rappers are not too happy with the changes Hammer has introduced. But Hammer takes it all in stride. "My competitors are dumbfounded and my success has confused them," said Hammer. "It's not the fact that anybody hates Hammer or hates his music–they hate the change."

Enough critics, reviewers, and fans enjoy Hammer's music, however, to make him one of the most popular entertainers of the 1990s. In 1991 alone, Hammer won many major awards. At that year's Grammy Awards, Hammer won best solo rap performance and best rhythm and blues song for "You Can't Touch This." *Please Hammer: The Movie* won best long-form video. At the People's Choice Awards, Hammer won best male performer.

And the list of awards kept growing. At the 1991 Billboard Magazine Music Awards, Hammer won for number one rap artist. He also won top honors at the 1990 MTV Video Music Awards and at the 1991 American Music Awards, Soul Train Awards, and Bay Area Music Awards.

In addition to these accomplishments, Hammer has been asked to endorse many popular items. By the end of 1990, he had signed up to do commercials for Pepsi and British Knights Sportswear. He's even done commercials for sneakers.

At the 1991 Grammy Awards, Hammer won three Grammys: Best Solo Rap Performance; Best Rhythm and Blues Song; and Best Long-Form Video.

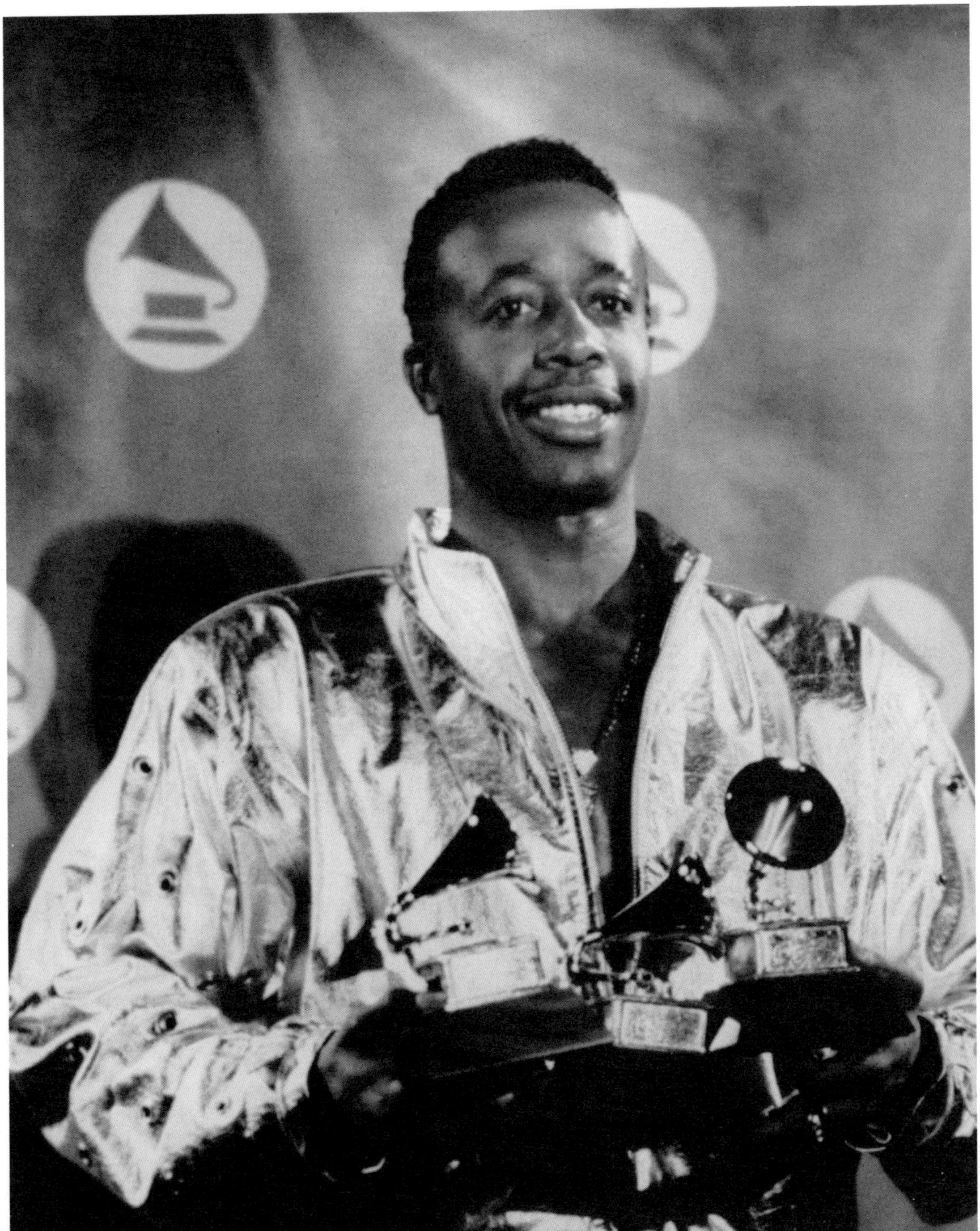

A TIRELESS BUSINESSMAN

Hammer loves to talk about his music, dancing, and record business, but he tries hard to keep his family life private. "As far as family goes, I don't make that public," he said in a September 1990 interview. "I have a two-year-old daughter named Akeiba Monique, and that's all I talk about. It is very hard for me to have a private life."

Today, Hammer and his daughter live in a five-bedroom home in Oakland. He works hard to set a good example for his daughter by writing music that doesn't just contain lyrics about dancing and having a good time. He tries to write songs that mean something or have strong, positive messages.

Hammer has used his money and influence to help other people. He's promised all the proceeds from his song "Help The Children" to a foundation for needy children. In March 1991, Hammer and Pepsi-Cola donated several thousand dollars to an organization in Australia that helps homeless youngsters. "I know how tough it is to grow up and I know how tough it is for parents," Hammer said. "That's why I'm willing to help."

Whenever he's back in Oakland, he visits the city's schools. "I make an active effort to remain a positive role model to kids," explained Hammer. "They need people to show them there's another way."

In October 1990, Hammer was invited back to Oakland Coliseum to throw out the ceremonial first ball before Game 3 of the American League play-offs. At that time, Hammer's old boss Charles Finley was in Europe negotiating a deal for the production of M.C. Hammer watches. "Charlie's working for me now," reported Hammer.

A few months later, in February 1991, Mattel Toys announced plans to create a twelve-inch M.C. Hammer doll. The doll is dressed in gold baggy pants and a short jacket. Hammer was on hand to introduce the doll to the public.

One job that keeps Hammer busy is that of chief executive officer of his record company. When Hammer signed with Capitol Records in 1988, he created Bustin' Records, an independent wing of Capitol. From his main office in Oakland, which is outfitted with a hot tub, Hammer fields acting offers and oversees his career. As part of his deal with Capitol Records, Hammer's company will produce albums for ten new groups over the next several years.

In early 1991, Hammer poses with young fans as he shows off the newest Mattell doll: M.C. Hammer dressed in gold baggy pants and a short jacket.

"My record company will be one of the strongest independent record companies in the music business," Hammer predicted. He hopes to work with artists who have styles similar to his. Bustin' Records' first release was "Go For It," the theme song for the movie *Rocky V*.

Hammer is determined to help others through his company. "Money isn't the ruling factor," he said. "What's important to me is that I'm in a position to set up my own music producing business and help others get established."

M.C.'S FUTURE PLANS

"Five years from now," predicted Hammer, "I hope to be at the point that I'm always coming out with material that is good and solid, and that you can look forward to, when Hammer is in town, seeing a good, fun, exciting show."

In addition to improving his music, Hammer hopes to one day act in a movie. He plans to write his own feature film. He has even begun work on another long-form video, which has already received over 100,000 advance orders.

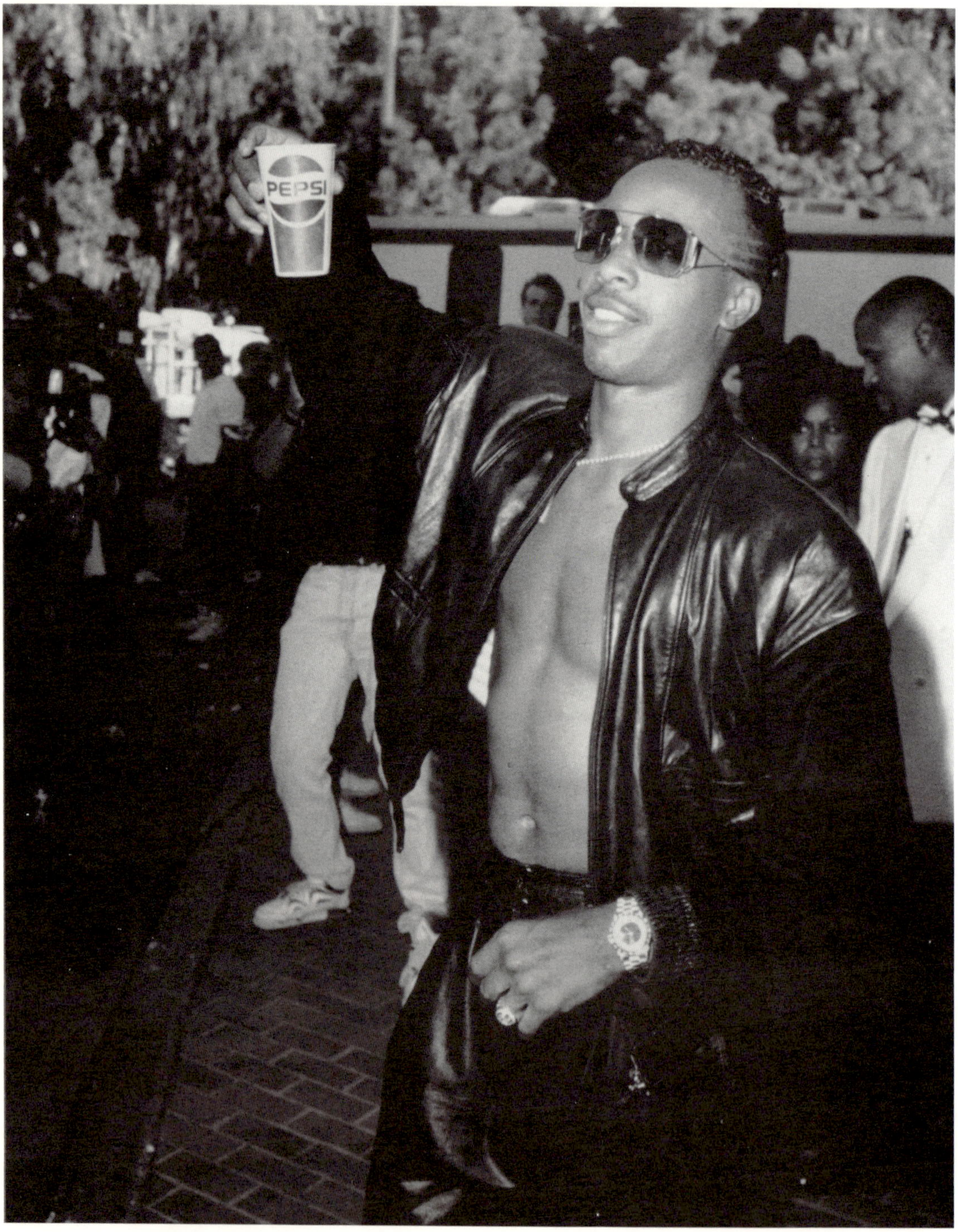
PEPSI

"I see myself as a film star at this point," said Hammer hopefully. "I'm not a singer-want-to-turn-movie-actor. I've always been an actor."

More endorsements are also sure to come in–along with new, more elaborate stage shows.

Whatever comes his way, Hammer will continue to remain positive, but not preachy. "I try to stay very real, very much in touch," he said. "I don't wanna be self-righteous. I never profess to be a perfect person because I'm not. I have my ups and downs and my faults and defaults."

One thing Hammer knows for sure is that the future will hold more rapping, more dancing, and more entertaining. "I don't plan on rapping when I'm fifty," he joked. "But I might rap when I'm forty."

Pepsi-Cola is only one of the products the versatile and talented Hammer sponsors.

M.C. HAMMER'S ADDRESS

You can write to M.C. Hammer at:

    M.C. Hammer
    c/o Capitol Records
    1750 N. Vine Street
    Hollywood, CA 90028